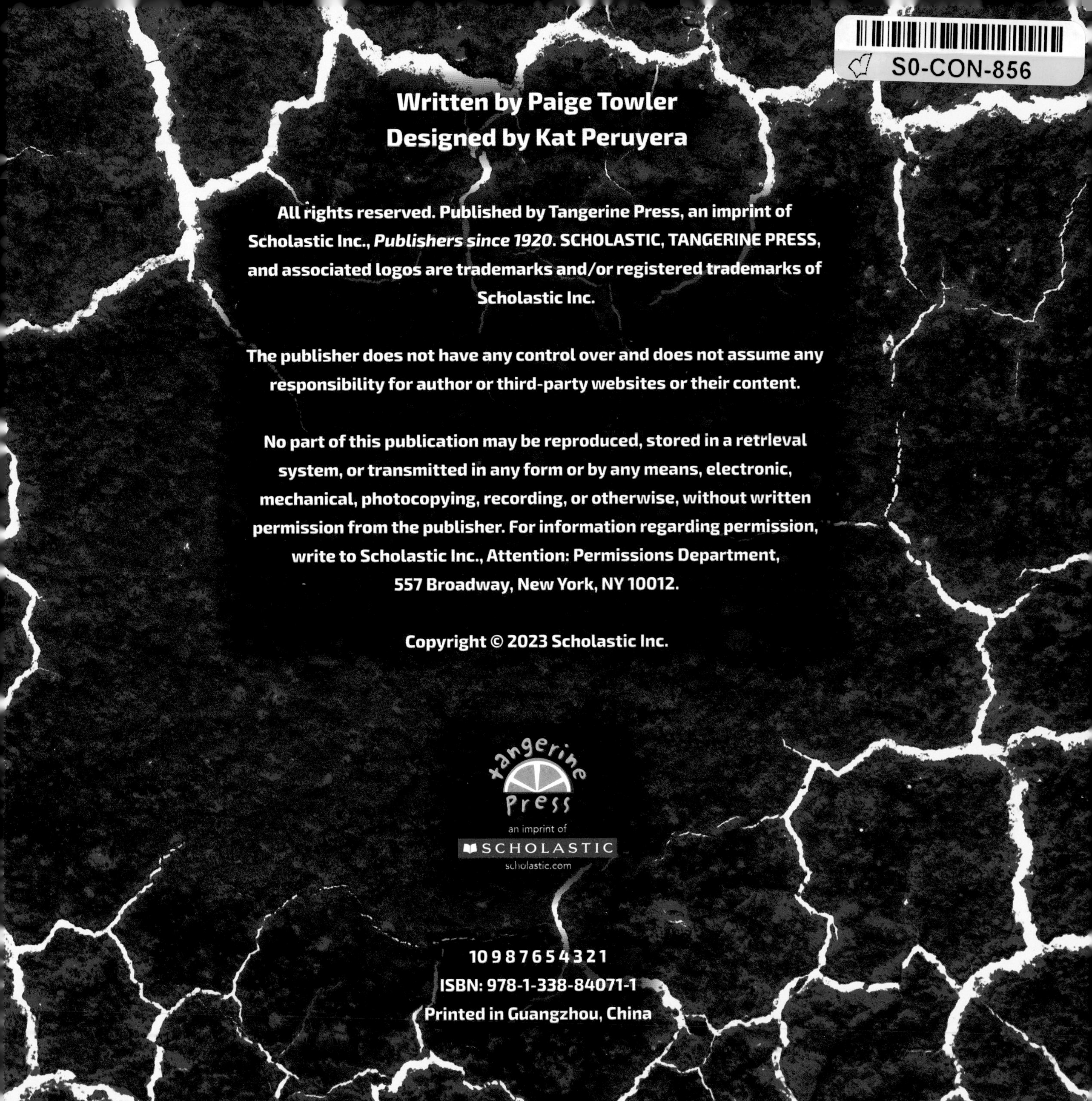

Written by Paige Towler
Designed by Kat Peruyera

tangerine press
an imprint of
SCHOLASTIC
scholastic.com

10 9 8 7 6 5 4 3 2 1
ISBN: 978-1-338-84071-1
Printed in Guangzhou, China

TABLE OF CONTENTS

EXPLORE THE AMAZING WORLD OF VOLCANOES

THERE ARE TONS OF FANTASTIC VOLCANOES AROUND THE WORLD!

But did you know that volcanoes come in many different shapes and sizes? There are volcanoes that look flat on the surface, volcanoes that remain inactive for hundreds of years, and even underwater volcanoes.

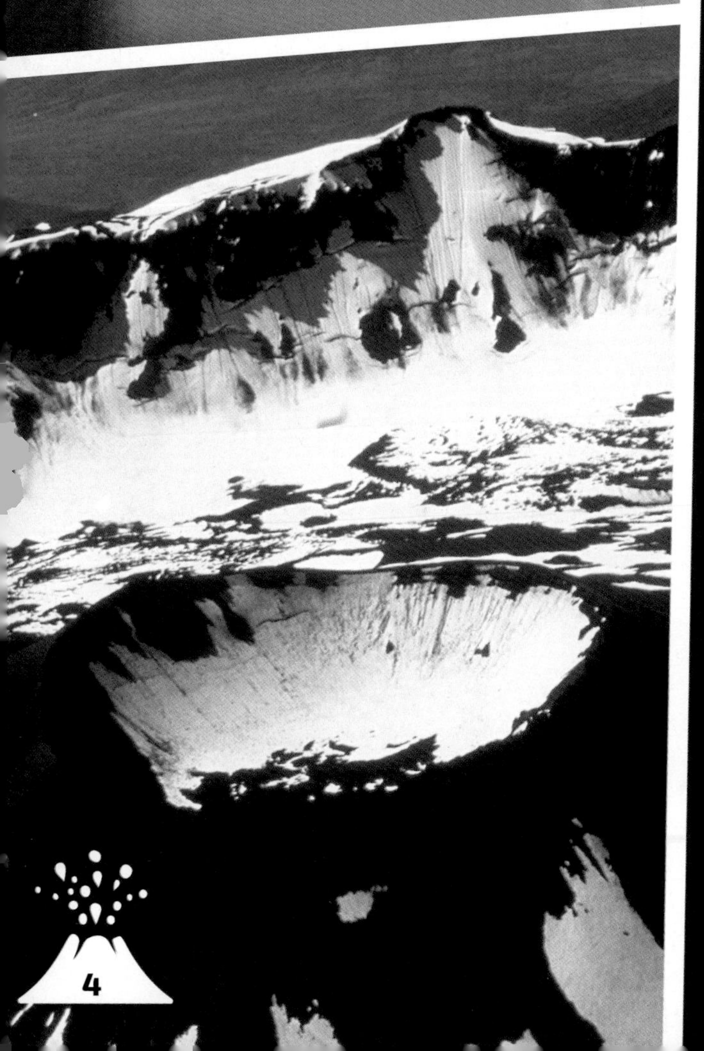

Get ready to learn what volcanoes are, how they form, why they **erupt**, and more. Plus, discover different types of volcanoes and what makes them special!

WHAT IS A VOLCANO?

A volcano is an opening deep into the surface of the earth. Gases, ash, and sometimes **lava** come out from deep inside the earth through these openings. When people think of volcanoes, they often think of mountains—but there are many kinds of volcanoes around the world.

The word "volcano" comes from the name for an ancient Roman god of fire, Vulcan.

VOLCANO STATS

Name: Augustine

Location: Alaska

Size: 4 ,108 feet (1,252 m) high

Type: **Lava Dome**

Last Erupted: 2006

Status: Active

A volcano is also known as a **vent**: an opening in the earth that pours out gases or lava.

A lava dome is a volcanic mound that is formed by cooling lava, which builds up over time.

EXPLOSIVE VOLCANOES

Pow! Sometimes volcanoes can erupt—or make huge explosions—that spew out hot lava. Lava is a very, very hot liquid-like form of rock. It is usually bright orange, due to its temperature. When it cools back down, lava turns back into solid rock.

Don't touch—liquid lava is usually between 1,300 and 2,200 degrees Fahrenheit (704.4 and 1,204.4 degrees C). That's up to about 10 times as hot as boiling water!

One type of volcanic rock is called **obsidian**. It can look like shiny, black glass.

Where lava is the hottest, it looks bright orange. In cooler places, it might look bright red or dark red, or even black.

Most people think of lava flowing down the side of a mountain; however, it originally comes from deep under the ground.

The earth is made of four main layers: crust, mantle, inner core, and outer core. When the middle or top layers get hot enough, they can melt into a **semi-liquid** form of **molten** rock. When this melted rock flows onto the surface of the earth, it is called *lava*. But when it is under the crust, it is known as **magma**.

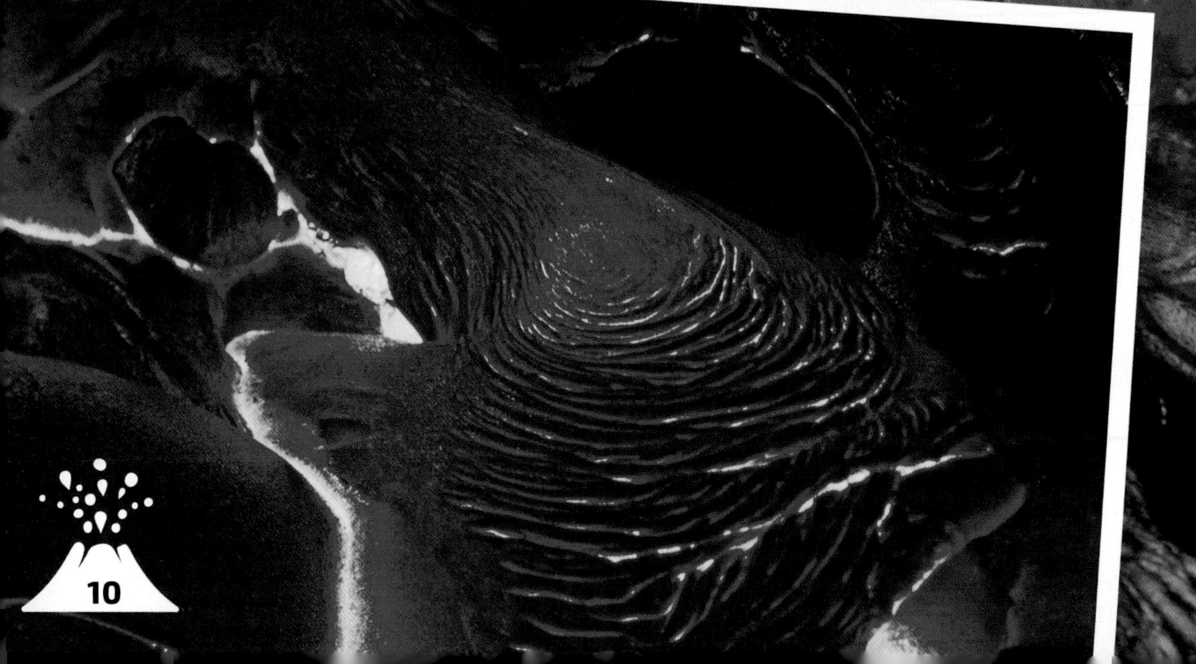

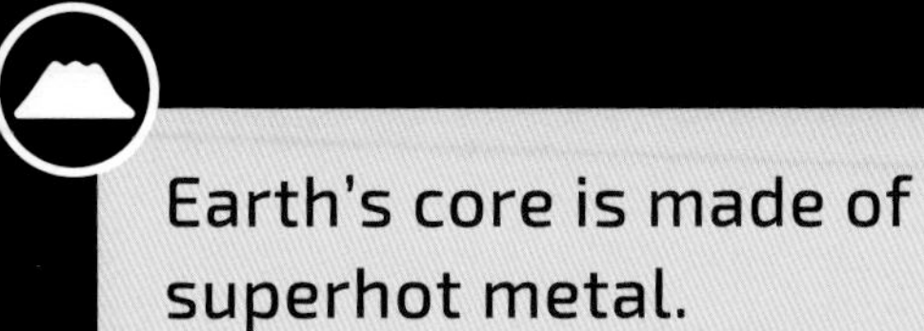

Earth's core is made of superhot metal.

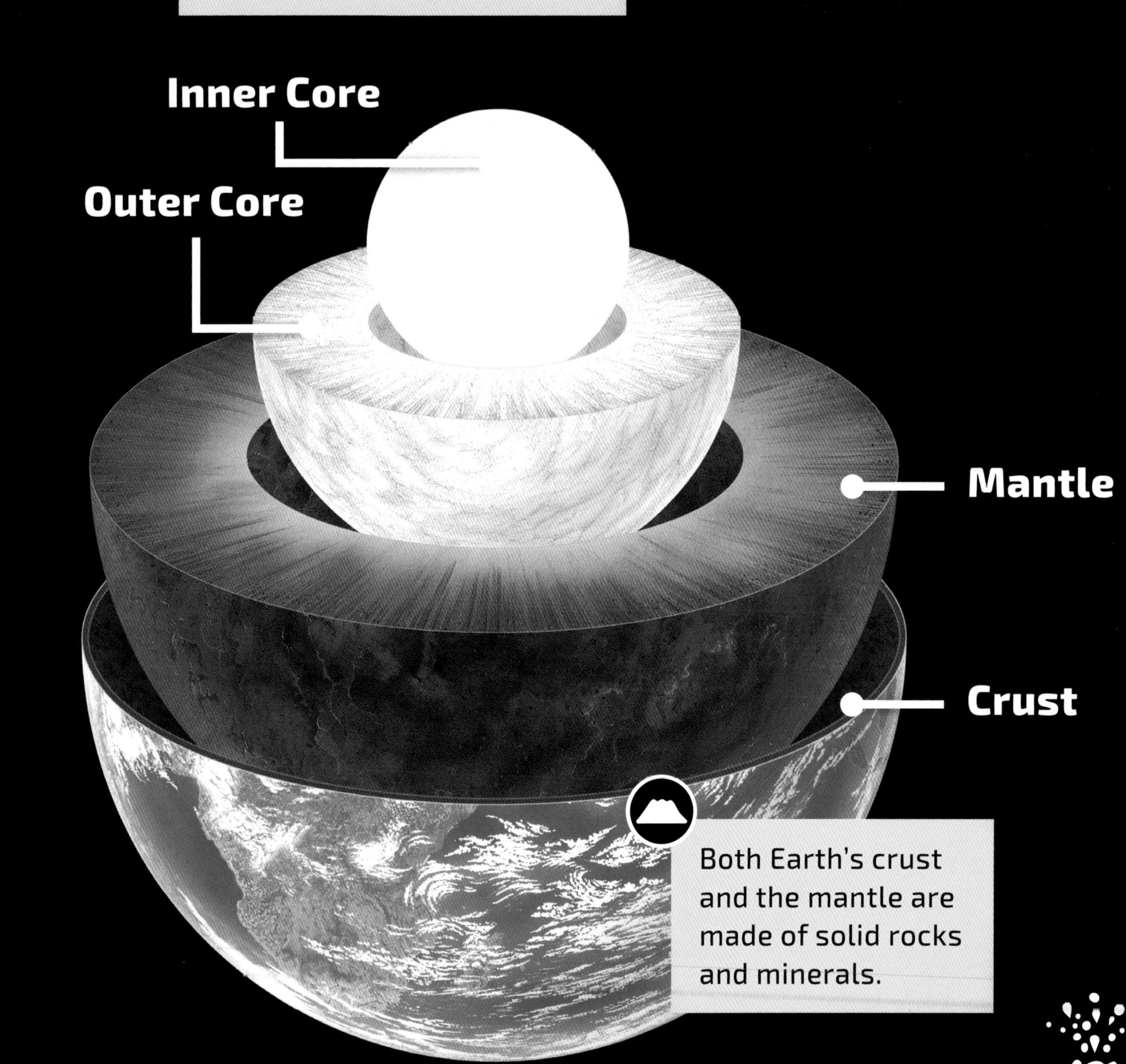

Both Earth's crust and the mantle are made of solid rocks and minerals.

EARTH ON THE MOVE

The earth's movements are why magma can travel from deep inside the earth and comes out as lava.

Earth's crust is broken up into large pieces, called *plates*. You normally can't feel it, but these plates are always moving around over Earth's inner layers—very, very slowly!

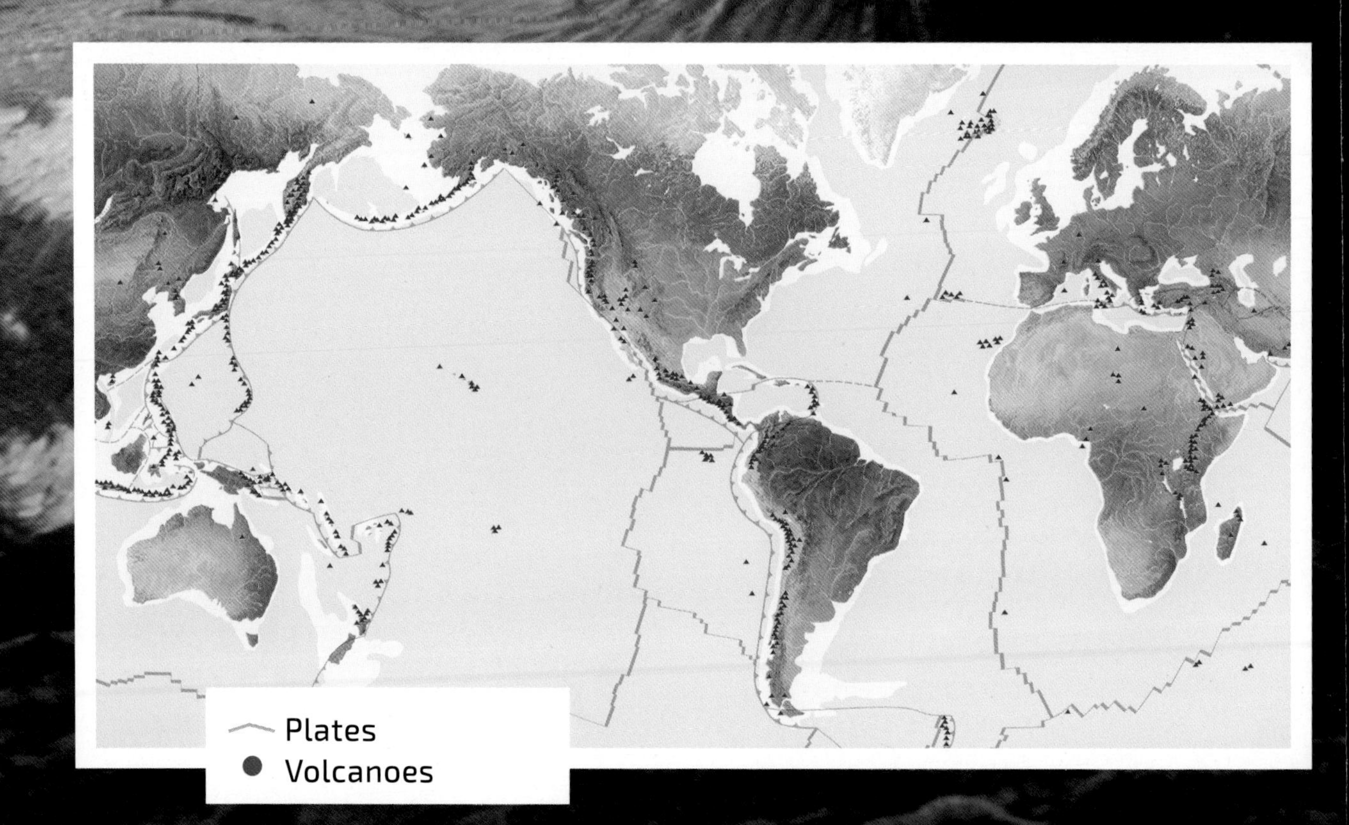

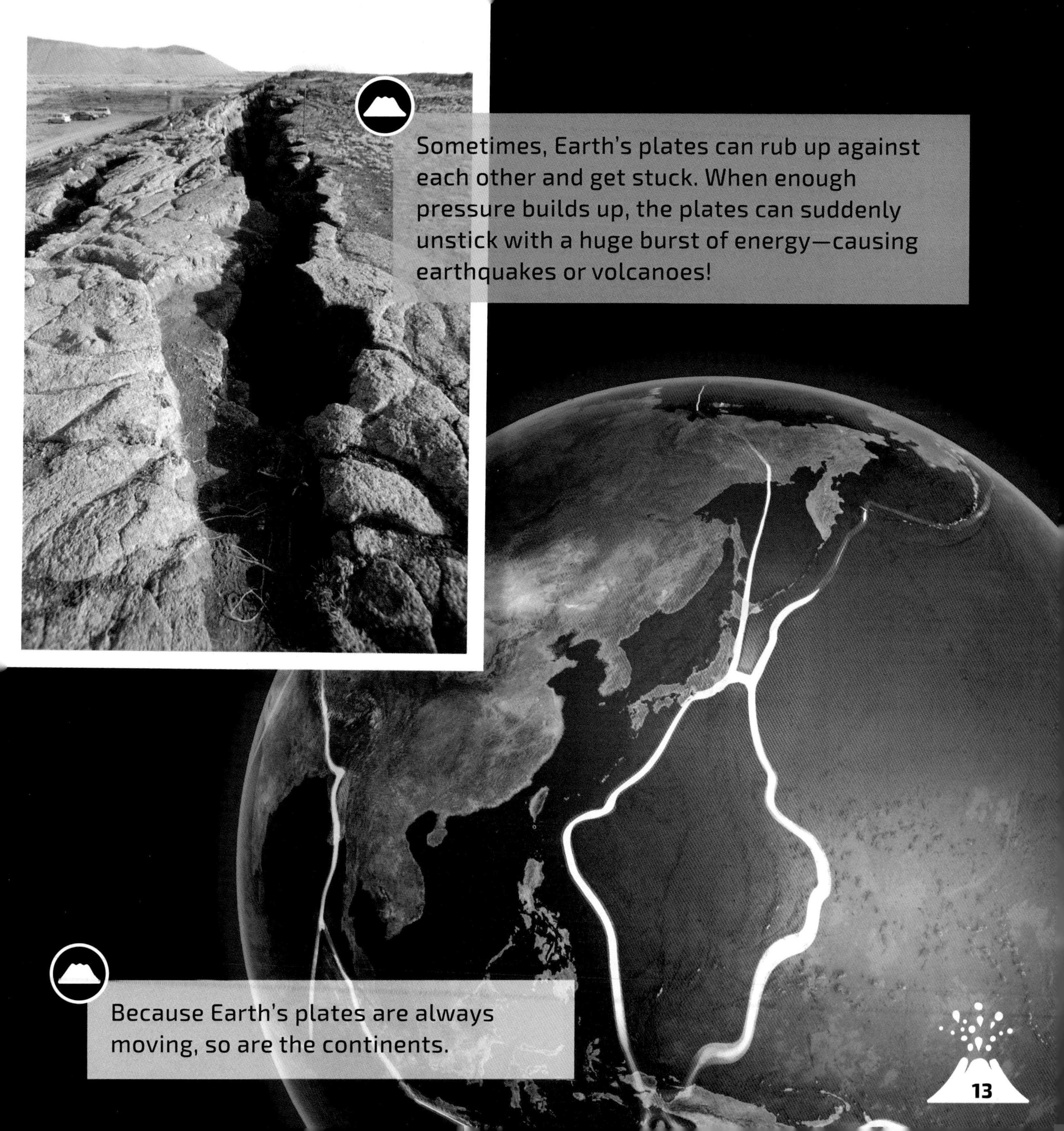

Sometimes, Earth's plates can rub up against each other and get stuck. When enough pressure builds up, the plates can suddenly unstick with a huge burst of energy—causing earthquakes or volcanoes!

Because Earth's plates are always moving, so are the continents.

FORMING SUBDUCTION VOLCANOES

Sometimes, two plates can slowly crash into each other. When this happens, the edge of one plate may be pushed down under the other. This process is called **subduction**, and it can create volcanoes!

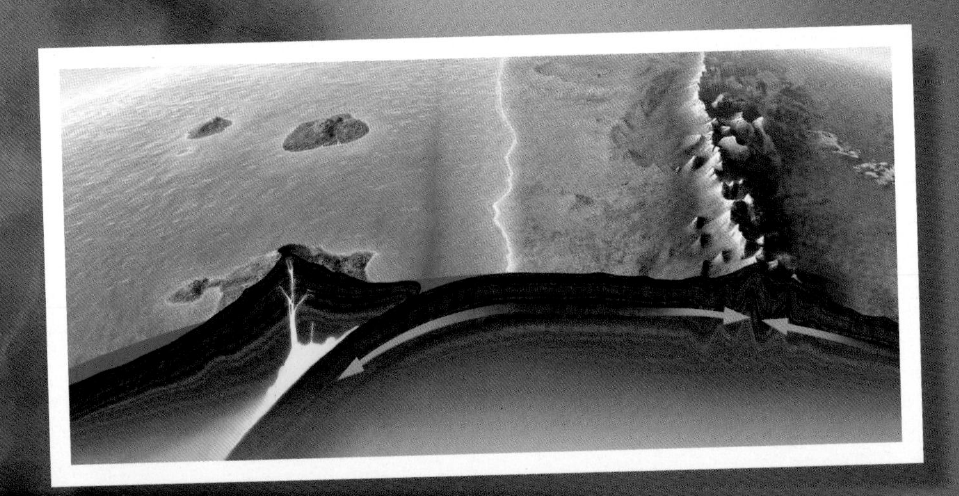

When one plate is pushed under another, the plate on top is often driven up to form mountains. Meanwhile, the plate that is pushed into the earth starts to melt, creating magma. Together, this makes a volcano.

Several plates along the Pacific Ocean create **subduction zones**. This area creates so many volcanoes that it is known as the **Ring of Fire**. There are more than 450 volcanoes here!

VOLCANO STATS

Name: Ring of Fire

Location: The Pacific Ocean

Size: 24,900 miles (40,072.7 km) long

Type: Many types

Last Erupted: Includes volcanoes currently erupting

Status: Active

MOUNT KATMAI

Mount Katmai is a subduction volcano. This volcano is located in Alaska along the Ring of Fire.

VOLCANO STATS

Name: Mount Katmai

Location: Katmai National Park and Preserve, Alaska

Size: 6,716 feet (2,047 m) high

Type: **Stratovolcano**

Last Erupted: June 6, 1912

Status: Active

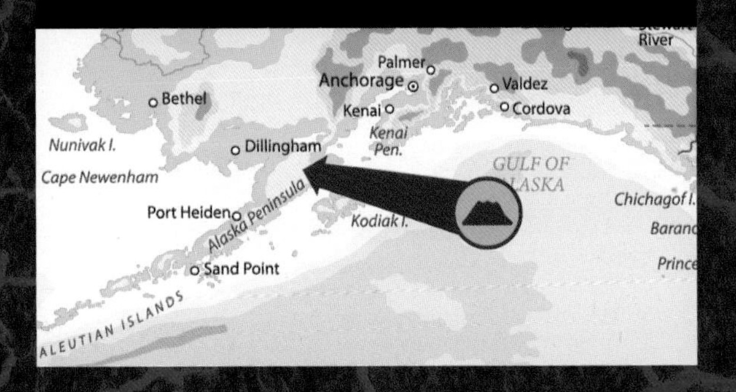

When Mount Katmai erupted in 1912, the top of the mountain fell in on itself. Today, the volcano holds a lake full of clear blue water.

When a volcano explodes, it can collapse in on itself and leave a **crater**. This crater is called a **caldera**.

Glaciers sometimes grow around the rim of Mount Katmai's lake.

FORMING DIVERGENCE VOLCANOES

Not all volcanoes form when plates collide. Other times, plates pull away from each other. This is called **divergence**.

Many places where plates diverge are under the ocean.

When two plates separate, magma pours out from inside the earth. The lava then cools into solid rock. Over time, this can build up into volcanoes! Volcanoes that form from plates separating under the ocean can create volcanic islands.

When lava meets ocean water, it cools very quickly. Sometimes, this lava cools into round or tube-shaped rocks, called **pillow lava**.

VALLES CALDERA

A surprising example of a volcano that is formed from diverging plates is located in New Mexico. A wide plain called the Valles Caldera is actually the caldera of a **supervolcano** that exploded more than one million years ago!

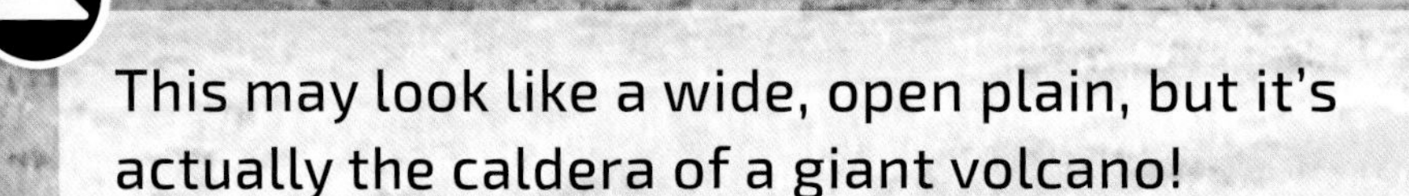

This may look like a wide, open plain, but it's actually the caldera of a giant volcano!

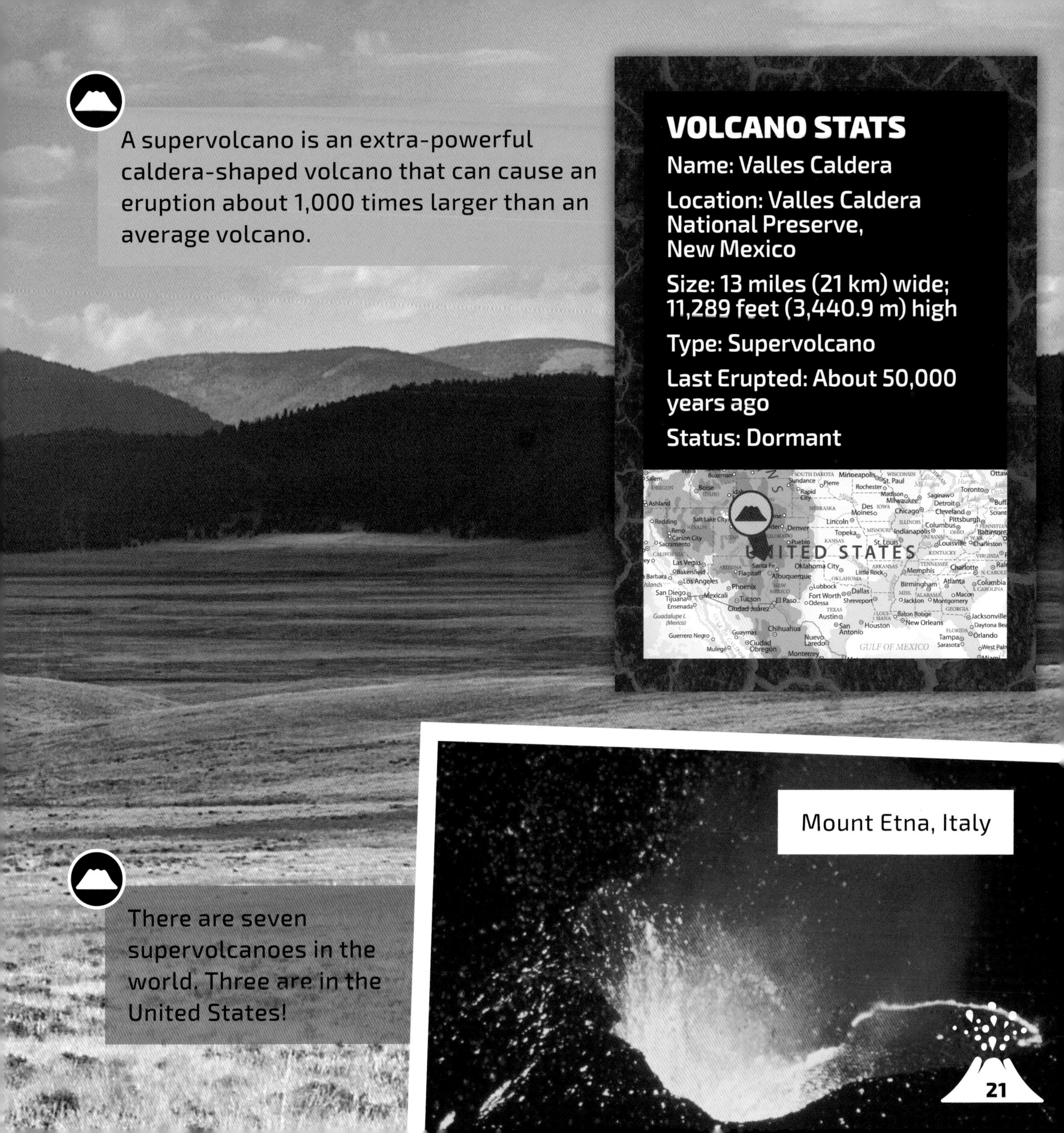

A supervolcano is an extra-powerful caldera-shaped volcano that can cause an eruption about 1,000 times larger than an average volcano.

VOLCANO STATS

Name: Valles Caldera

Location: Valles Caldera National Preserve, New Mexico

Size: 13 miles (21 km) wide; 11,289 feet (3,440.9 m) high

Type: Supervolcano

Last Erupted: About 50,000 years ago

Status: Dormant

Mount Etna, Italy

There are seven supervolcanoes in the world. Three are in the United States!

FORMING HOTSPOT VOLCANOES

Not all volcanoes form at the edges of plates. Instead, they form at places called **hotspots**.

Some parts of the earth's mantle are extra hot. In these places, heat streams up to the surface, creating and sending magma along with it—and forming volcanoes!

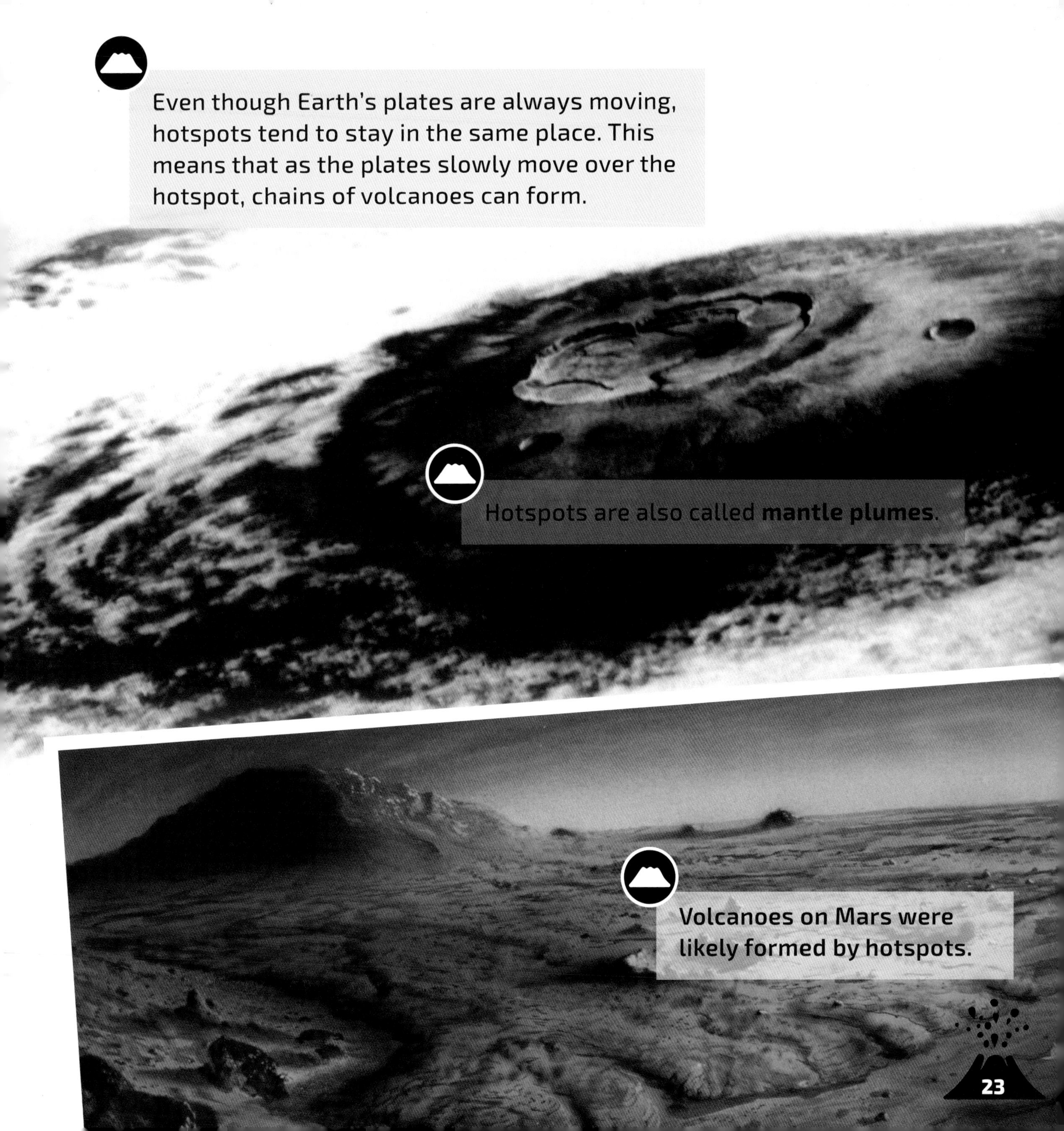

Even though Earth's plates are always moving, hotspots tend to stay in the same place. This means that as the plates slowly move over the hotspot, chains of volcanoes can form.

Hotspots are also called **mantle plumes**.

Volcanoes on Mars were likely formed by hotspots.

KO'OLAU

In the state of Hawaii, there are many volcanoes created by a hotspot. In fact, all of the 137 islands that make up the state of Hawaii were formed by a hotspot!

Ko'olau is one of the volcanoes in Hawaii. Its enormous crater was formed during an eruption more than 300,000 years ago.

According to tradition, the Hawaiian goddess Pele is a volcano goddess.

VOLCANO STATS

Name: Ko'olau
Location: Hawaii
Size: 3,087 feet (940.9 m) high; 3,520 feet (1,072.9 m) across
Type: Shield
Last Erupted: More than 11,000 years ago
Status: Dormant

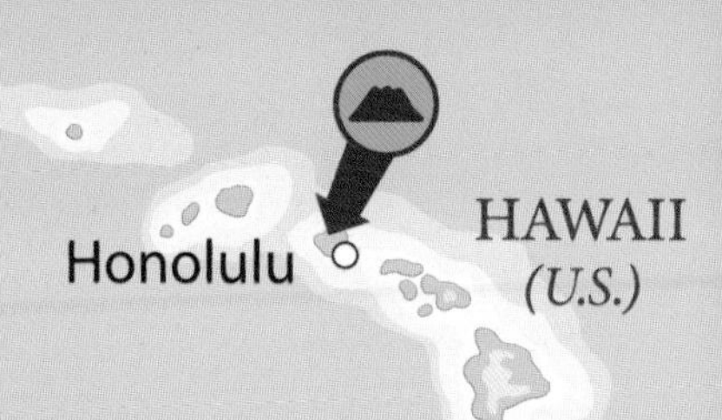

Hotspots can be on land or in the ocean.

CINDER CONE VOLCANOES

Scientists usually organize volcanoes into types. The three most common types are **cinder cone volcanoes**, stratovolcanoes, and **shield volcanoes**. Cinder cones are the most common type of all.

When most people imagine a volcano, they often think of a cone-shaped mountain—a cinder cone volcano! Cinder cones form from small pieces of lava that are thrown from the volcano's vent and build up over time.

Wizard Island sits inside Crater Lake, a caldera formed by the much larger, older volcano called Mount Mazama.

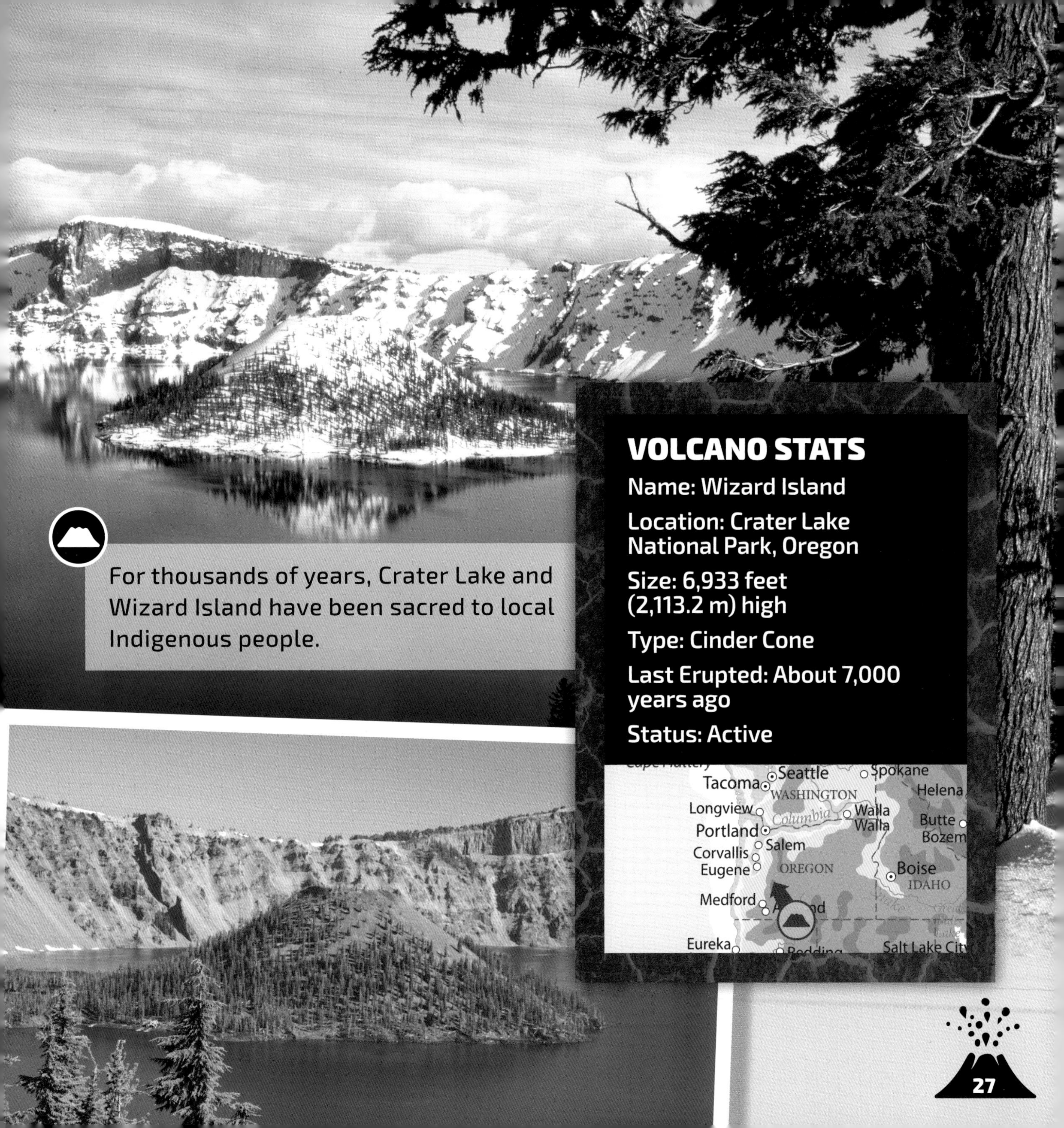

For thousands of years, Crater Lake and Wizard Island have been sacred to local Indigenous people.

VOLCANO STATS

Name: Wizard Island

Location: Crater Lake National Park, Oregon

Size: 6,933 feet (2,113.2 m) high

Type: Cinder Cone

Last Erupted: About 7,000 years ago

Status: Active

STRATOVOLCANOES

A stratovolcano is a volcano made up of layers of ash and cooled lava. These volcanoes usually have gentle **slopes** and steep **peaks**.

Stratovolcanoes often cause huge eruptions. One example of a stratovolcano is Mount St. Helens in Washington.

VOLCANO STATS

Name: Mount St. Helens

Location: Washington

Size: 8,363 feet (2,549 m) high

Type: Stratovolcano

Last Erupted: 2008

Status: Active

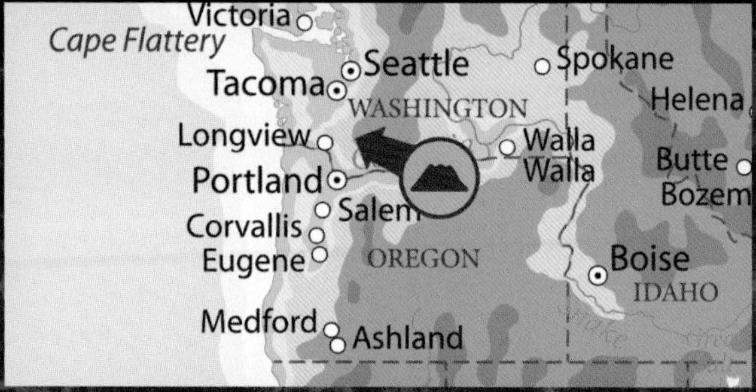

In 1980, Mount St. Helens erupted with so much force that its peak exploded.

Stratovolcanoes are also called **composite volcanoes**.

SHIELD VOLCANOES

Shield volcanoes are wide volcanoes. They form from lava that flows down the volcanoes' sides. These volcanoes can grow very big.

Mauna Loa is one of the volcanoes in Hawaii. It is one of the largest volcanoes in the world!

Mauna Loa means "long mountain" in Hawaiian.

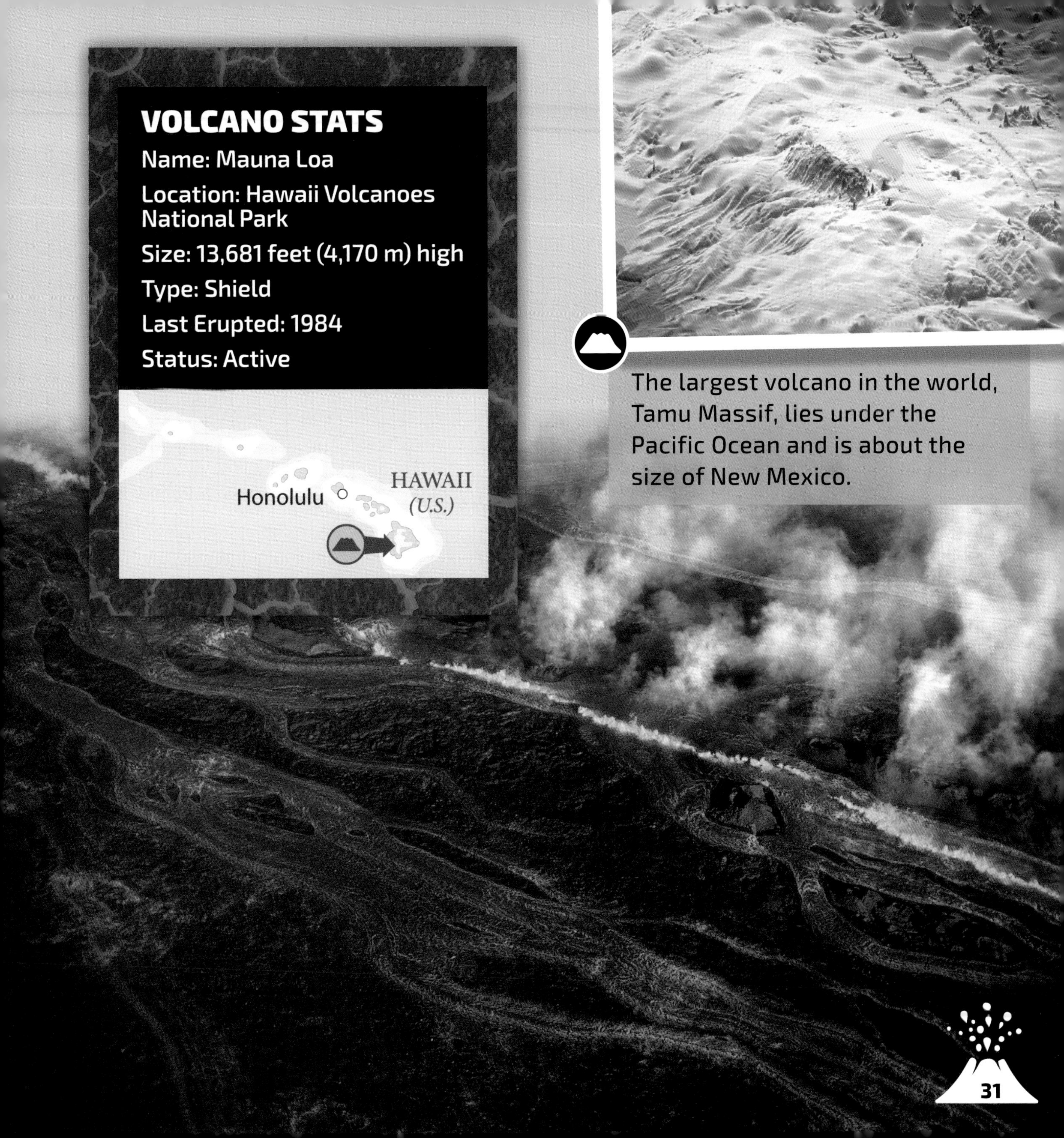

VOLCANO STATS

Name: Mauna Loa

Location: Hawaii Volcanoes National Park

Size: 13,681 feet (4,170 m) high

Type: Shield

Last Erupted: 1984

Status: Active

The largest volcano in the world, Tamu Massif, lies under the Pacific Ocean and is about the size of New Mexico.

OTHER TYPES OF VOLCANOES

There are other types of volcanoes that are less common too. One type is a **fissure volcano**: a long crack in the earth's surface where lava erupts. These are also known as fissure vents.

Fagradalsfjall in Iceland is made up, in part, of many fissure volcanoes.

A long fissure volcano erupting all at once is sometimes referred to as a *curtain of fire*.

VOLCANO STATS

Name: Fagradalsfjall

Location: Iceland

Size: 1,263 feet (385 m) long

Type: Fissure

Last Erupted: Currently has active lava flows

Status: Active

The lumps of lava that spurt out of a fissure volcano and hit the ground are known as *spatter*.

SLEEPING VOLCANOES

No matter what type it is, a volcano will have one of three different statuses: active, dormant, or extinct.

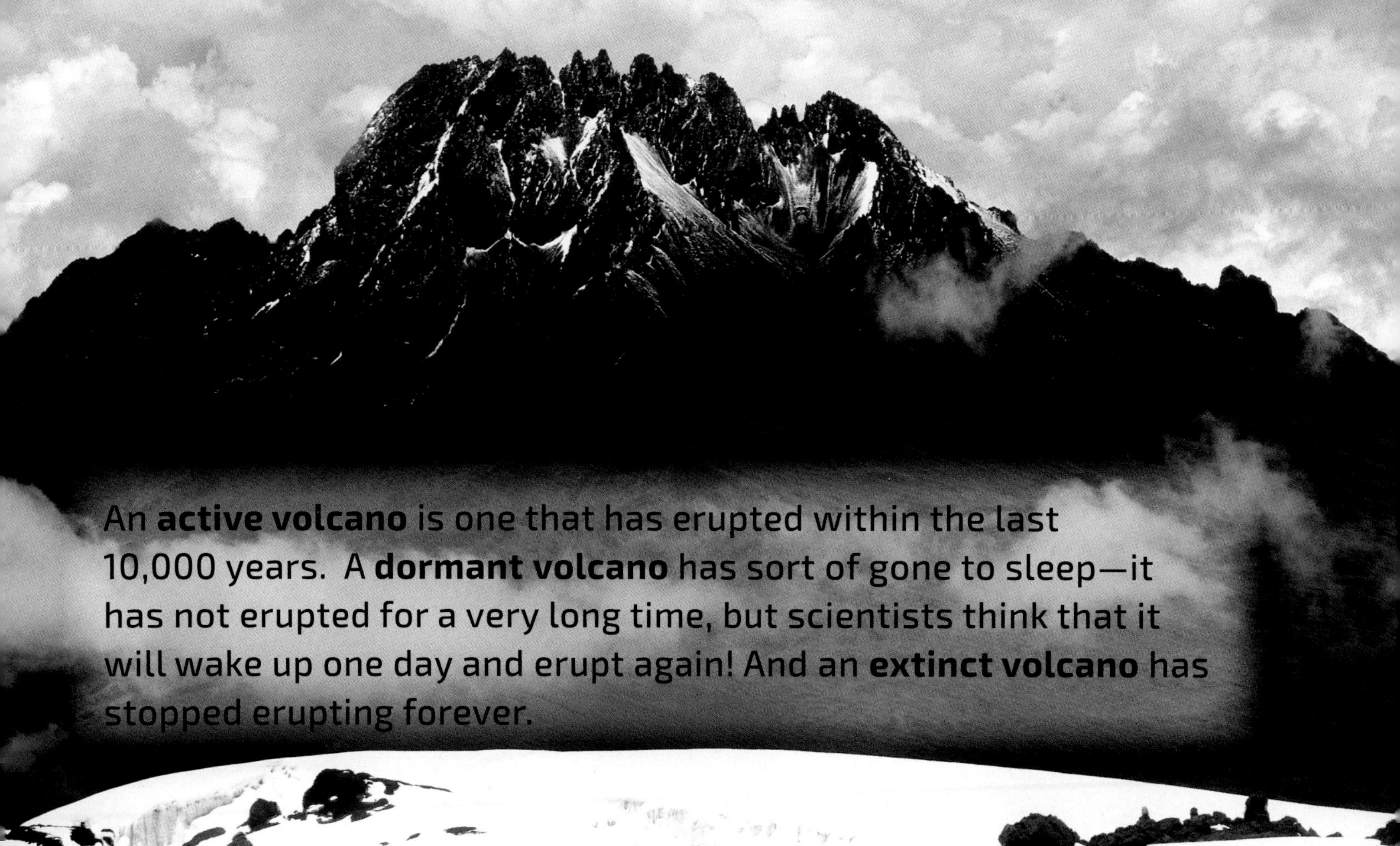

An **active volcano** is one that has erupted within the last 10,000 years. A **dormant volcano** has sort of gone to sleep—it has not erupted for a very long time, but scientists think that it will wake up one day and erupt again! And an **extinct volcano** has stopped erupting forever.

VOLCANO STATS

Name: Mount Kilimanjaro

Location: Mount Kilimanjaro National Park, Tanzania

Size: 19,340 feet (5,894.8 m) high

Type: Stratovolcano

Last Erupted: About 200,000 years ago

Status: Dormant

There aren't any active volcanoes in the Eastern part of the United States, but there are extinct ones: Scientists discovered two ancient, extinct volcanoes—named Jackson and Midnight—buried deep under the earth in parts of Mississippi.

OOZING VOLCANOES

Deep within an active volcano, magma and gas build up over time. In some volcanoes, the magma and gas build so quickly that the volcano oozes lava all the time!

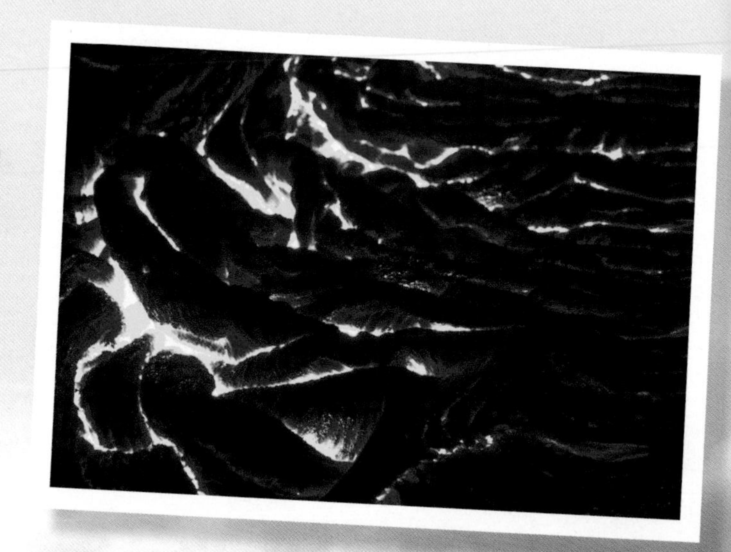

This happens when the lava is very liquid-like and can easily escape to the surface. Kīlauea is a volcano that has lots of lava flows. It is the most active volcano in the world!

In Hawaiian, Kīlauea means "much spreading."

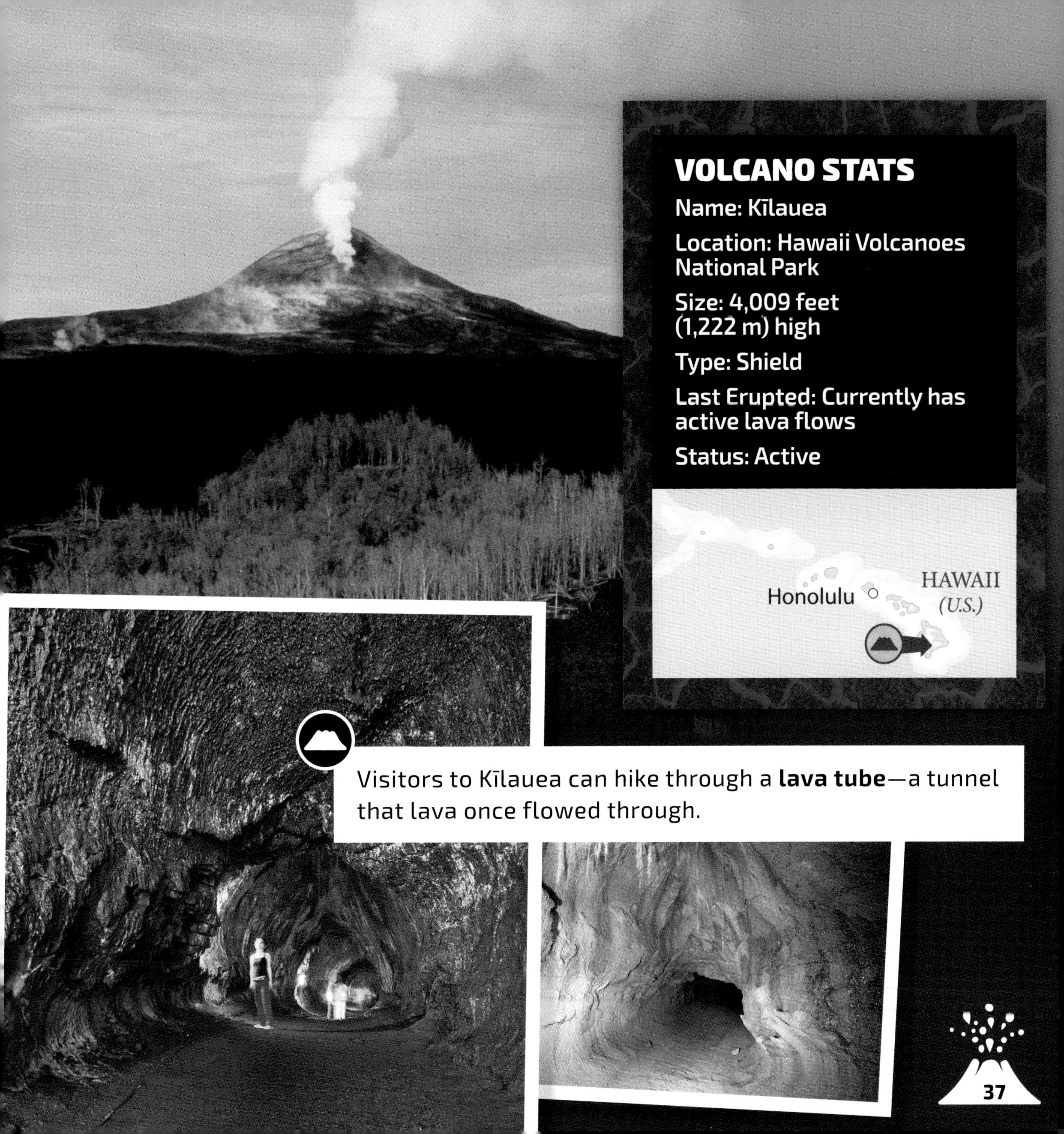

VOLCANO STATS

Name: Kīlauea

Location: Hawaii Volcanoes National Park

Size: 4,009 feet (1,222 m) high

Type: Shield

Last Erupted: Currently has active lava flows

Status: Active

Visitors to Kīlauea can hike through a **lava tube**—a tunnel that lava once flowed through.

WHY DO VOLCANOES ERUPT?

In some volcanoes, the gas and magma cannot escape so easily. If the magma is sticky and thick, or if the volcano's vents get blocked, gas can't get out.

When this happens, pressure grows and grows . . . until it erupts out of the volcano!

Scientists say there are about 1,500 active volcanoes in the world.

EXPLOSIVE ERUPTIONS

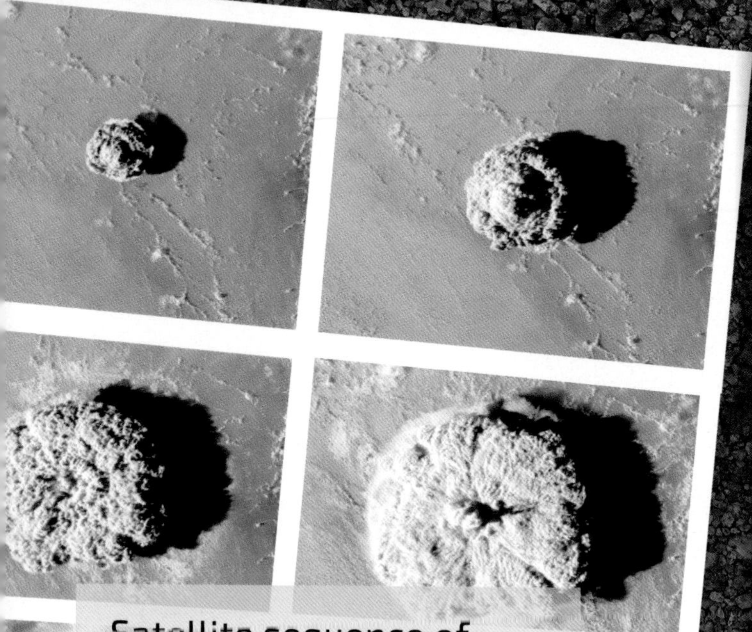

Satellite sequence of Hunga Tonga-Hunga Ha'apai eruption

An exploding volcano can be a very powerful event. When the pressure from the buildup of gas is too much, it explodes outward, forcing lava, rocks, ash, and gas into the air.

These volcanic explosions can cover huge areas in ash, create landslides, and send out giant waves of heat. In 2022, the volcano Hunga Tonga-Hunga Ha'apai erupted with an explosion so large it created deadly giant waves, called *tsunamis*, which reached thousands of miles away.

Volcanic eruptions can create lightning.

It takes about 1,000 years for Hunga Tonga-Hunga Ha'apai to fill with enough magma and pressure to erupt.

The Hunga Tonga-Hunga Ha'apai eruption was so big and loud that it was heard 1,430 miles (2,301.4 km) away in New Zealand.

VOLCANO STATS

Name: Hunga Tonga-Hunga Ha'apai

Location: Tonga

Size: 374 feet (114 m) high above the water

Type: Submarine

Last Erupted: 2022

Status: Active

Hunga Tonga-Hunga Ha'apai is a submarine volcano: a volcano that is mostly underwater.

AFTER THE ERUPTION

Volcanoes are powerful and can cause a lot of damage.

But volcanoes are also powerful at creating! Cooled lava creates new land, the eruptions and ash bring needed minerals to plants and animals, and the heat from volcanoes can be used to create electricity!

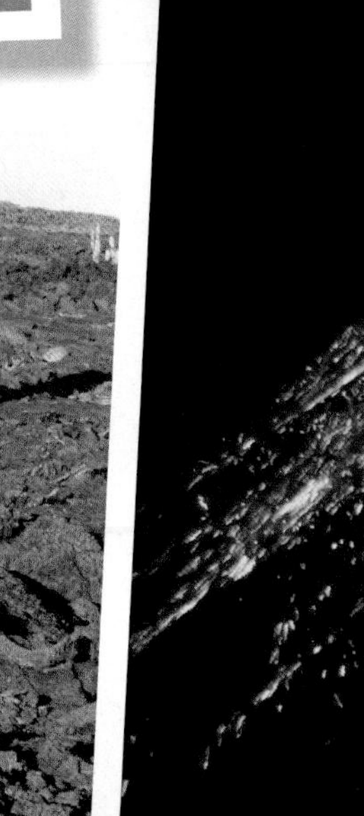

Volcanic activity has created more than 80 percent of the earth's surface.

LIVING WITH VOLCANOES

Volcanoes are very powerful, but they are also really wonderful. Today, scientists have lots of tools that help them study volcanoes to learn more about them and keep people safe.

VOLCANO STATS

Name: Mount Etna

Location: Parco dell'Etna, Italy

Size: 10,892 feet (3,319.9 m) high

Type: Stratovolcano

Last Erupted: Currently has active lava flows

Status: Active

Scientists can help track the amount of energy or gas a volcano puts out—also called the **seismic activity**—to help predict when an eruption might occur.

Scientists help keep people who live near volcanoes—like those near Mount Etna—safe. They also learn how volcanoes can help humans and the environment.

Experts can help teach people about the safest practices to use when they are living near a volcano.

GLOSSARY

Active volcano: a volcano that has erupted within the last 10,000 years

Caldera: a large sunken area that forms when a volcano erupts and collapses

Cinder cone volcano: a cone-shaped volcano created by a buildup of lava

Composite volcano: a volcano made from layers of ash and lava; another word for stratovolcano

Crater: a large, bowl-shaped sunken area in the ground

Divergence: the act of two tectonic plates pulling away from each other

Dormant volcano: a volcano that has not erupted for a very long time, but may erupt again

Erupt: to put out lava and ash

Extinct volcano: a volcano that has not erupted for a very long time and is not likely to erupt ever again

Fissure volcano: a long crack in the earth that erupts lava, ash, and gas

Glacier: a large, slowly moving body of ice

Hotspot: an area of the earth where heat rise from the mantle through the crust

Lava: hot molten rock that pours onto the surface of the earth

Lava dome: a mound of lava that forms over time

Lava tube: a tunnel formed by the flow of lava

Magma: very hot liquid and semi-liquid rock beneath Earth's surface

Mantle plume: a hotspot

Molten: heated to the point of melting to a semi-liquid state

Obsidian: a glass-like, volcanic rock

Peak: the highest part of a mountain

Pillow lava: lava that has cooled from contact with water, forming a rounded shape

Ring of Fire: an area around the Pacific Ocean of high volcanic formation

Seismic activity: the rate and size of earthquakes experienced over a period of time

Semi-liquid: having nearly liquid qualities

Shield volcano: a wide volcano created by cooling lava flows

Slope: the angle of a surface where one end is higher than the other

Stratovolcano: a volcano made from layers of ash and lava; another word for composite volcano

Subduction: the downward motion of one tectonic plate beneath another

Subduction zone: an area in which subduction is happening

Supervolcano: an extra-large, caldera-shaped volcano

Vent: another word for volcano; an opening in Earth's surface from which ash, gas, and lava come out

METRIC TABLE

The metric system is a system of measurements. It is used in many parts of the world. It is also used by all scientists, no matter where they live. Here are some common abbreviations and conversions for metric measurements used in this book.

TEMPERATURE

C = Celsius

(1 degree Celsius = 33.8 degrees Fahrenheit)

DISTANCE

km = kilometers (1 kilometer = .6 mile)

m = meters (1 meter = 3.3 feet)